Damn Particles

PHYSICS CARTOONS

by **SIDNEY HARRIS**

Sidney Harris,
  DAMN PARTICLES!
  Physics Caroons

ISBN: 978-0-9890685-2-9

Many of the cartoons in this book have been previously published
and appeared in the following magazines: *American Scientist,
Chronicle of Higher Education, Discover, Fantasy and Science Fiction,
Johns Hopkins Magazine, Physics Today, Saturday Review, Science,
Science 80, Scientific American, The New Yorker, Today's Chemist*

Sidney Harris
Box 1980, Federal Station
New Haven, CT 06521
USA

Or

sharris777@aol.com
ScienceCartoonsPlus.com

Logo Design by Martha Bradshaw

BOOKS BY THE SAME AUTHOR:

FOREWORD

The odd couple of physics and cartoons might correspond to the Mary Poppins suggestion: "Just a spoonful of sugar helps the medicine go down . . .In a most delightful way."

That's probably why my writing partner, Charlie Wynn, and I approached Sidney Harris 20+ years ago to provide cartoon commentary for our book *"The Five Biggest Ideas in Science."* Many books later, it has become clear that there is a lot more to the story.

Physics' medicinal aspects are often related to its use of mathematics, a foreign language to many, while Sidney's cartoons are sugary, with their simplified appearance and occasional pithy captions. Sidney's talent at encapsulating complex ideas with an economy of inked lines that flow from his pen are not only impressive, they require a substantially different skill set than mathematical manipulations. Who can ever forget his cartoon showing Einstein, standing at a blackboard, staring at crossed-out equations $E = ma2$ and $E = mb2$? Sidney Harris cartoons not only comment on physics' efforts to understand the universe, but the human aspect of the attempts.

This Brooklyn Dodgers fan (don't mention the Yankees to him) is no one-trick pony, either. Besides science, Sidney Harris cartoons comment on math, business, medicine, law, the environment, psychology, global warming, food, and animals. More than 600 of his cartoons have appeared in the *American Scientist*, plus many other magazines, especially *The New Yorker*. He was elected an honorary member of Sigma Xi, the scientific research honor society, and an exhibit of his cartoons and paintings has been touring the country's museums since 1985. Other cartoon books include *Einstein Atomized: More Science Cartoons, Chalk Up Another One, Can't You Guys Read? Cartoons on Academia, Stress Test: Cartoons on Medicine, 101 Funny Things About Global Warming, So Sue Me! Cartoons on the Law, Chicken Soup and Other Medical Matters, There Goes the Neighborhood: Cartoons on the Environment, Freudian Slips, Einstein Simplified, From Personal Ads to Cloning Labs, You Want Proof? I'll Give You Proof, and 49 Dogs, 36 Cats & A Platypus*. More than 35,000 cartoons make up the prodigious output of this Brooklyn College English major, who has, in his own words, ". . . never had a job."

Although I've seen almost all of Sidney's past efforts, this collection contains loads of new ones that occasioned laugh out loud moments. He has a unique talent that tickles the viewer to think outside the box because of his unusual perspective.

Arthur W. Wiggins
Distinguished Professor Emeritus of Physics, Oakland Community College

"QUARKS. NEUTRINOS. MESONS. ALL THOSE DAMN PARTICLES YOU CAN'T SEE. THAT'S WHAT DROVE ME TO DRINK. BUT NOW I CAN SEE THEM."

"EVEN I FIND IT HARD TO BELIEVE SOME OF THE THINGS I'VE BEEN COMING UP WITH."

"JUST BECAUSE THE ALIGNMENT OF THE SYSTEM WASN'T MAINTAINED DURING RAPID SAMPLE EXCHANGE, IT'S NO THREAT TO YOUR MANHOOD."

"IT APPEARS YOU'VE FOUND
A NEW PARTICLE."

Papa Newton reading some bedtime equations to little Isaac

JOHN VON NEUMANN, AFTER COMPLETING HIS BOOK, 'THEORY OF SELF-REPLICATING AUTOMATA'

HARNESSING THE BLACK HOLE

"HE WAS WORKING ON A THEORY OF ENTROPY, AND DEVELOPED A SEVERE CASE OF IT HIMSELF."

EINSTEIN SIMPLIFIED

I apply for a grant, and all they give me is a big ball and a small ball. Here's what I think of them.
S. Harris

"IT'S DOWN. THE PARTICLES ARE SUPPOSED TO COLLIDE WITH OTHER PARTICLES, NOT WITH THE PHYSICISTS."

科学 俳句

未知物仍然在蒙蒙的
影子后面微笑.
如何才能发现它?有何目的?

PHYSICS HAIKU

THE UNKNOWN STILL
SMILES
BEHIND A HAZY SHADOW.
HOW TO FIND IT. WHY?

HIGH-GRAVITY BASEBALL

"I AM CONVINCED THAT GOD DOES NOT PLAY CARDS WITH THE UNIVERSE."

EINSTEIN CONDUCTS A THOUGHT-EXPERIMENT

"IT GOES ON TO SAY 'WE USE PETROLEUM TO LIGHT OUR LAMPS, AND TO HEAT OUR FOOD, BUT WE CAN'T SEEM TO COME UP WITH AN INTERNAL COMBUSTION ENGINE'."

SCIENCE AND SOCIETY—1923

SCHRÖDINGER, PAULI, HEISENBERG, PLANCK,
BOHR, CURIE, EINSTEIN, KIKI AND BUD VANDERVELT

"WHAT I MISS MOST IS THE MOOING."

ALBERT EINSTEIN TELLS A JOKE...

GALILEO, J. ROBERT OPPENHEIMER, A RUSSIAN COSMONAUT, TWO ELECTRICIANS, MADAME CURIE, JASCHA HEIFETZ, THREE SOPRANOS, A CHIMPANZEE AND NIELS BOHR MEET IN SIGMUND FREUD'S WAITING ROOM...

"MY RESEARCH COVERS TWO FIELDS: THE BEHAVIOR OF MATTER UNDER HIGH PRESSURE, AND THE BEHAVIOR OF SCIENTISTS UNDER HIGH PRESSURE."

"YOU BOTH HAVE SOMETHING IN COMMON. DR. DAVIS HAS DISCOVERED A PARTICLE WHICH NOBODY HAS SEEN, AND PROF. HIGBE HAS DISCOVERED A GALAXY WHICH NOBODY HAS SEEN."

"ACCORDING TO THE SPACE-TIME CONTINUUM, OUR LUNCH SHOULD HAVE BEEN HERE SEVENTEEN MINUTES AGO."

"BELIEVE ME, IF YOU LIVED AROUND HERE AS LONG AS I HAVE, YOU'D SEE PLENTY OF SNOWFLAKES THAT LOOK ALIKE."

ISAAC NEWTON, PURSUING HIS STUDY OF THE LAWS OF MOTION, DISCOVERS THAT A THROWN BALL CAN CURVE...
But no one will ever care about this.
S. Harris

"MY SECRET AMBITION IS TO FIND A NEW ELEMENT, BUT TO TELL THE TRUTH, I DON'T KNOW WHERE TO LOOK."

A NATION OF EINSTEINS

"I GUESS THERE'LL ALWAYS BE A GAP BETWEEN SCIENCE AND TECHNOLOGY."

35초 동안 사는 물체와 일하는 것

쿼크박사 언제 먹기를 시작해서 언제 다 먹었는지를 모르는 햄버거에 익숙 해진 사람들의 생활 모습

"THIS IS SUPPOSED TO BE THE CLEAN ROOM, SO WE'LL GIVE IT AN ADDITIONAL FIVE MINUTES."

"WE'VE PROVEN, WITHOUT A DOUBT, THAT THIS PARTICLE HAS A NEGATIVE CHARGE. UNFORTUNATELY AN ACCELERATOR IN SWITZERLAND HAS PROVEN, WITHOUT DOUBT, THAT IT HAS A POSITIVE CHARGE."

TEN COMMANDMENTS AND THE PERIODIC TABLE — ALL WE HAVE TO KNOW!
S. harris

THEORETICAL PHYSICIST WITH POWERS OF ESP STEALING A THOUGHT EXPERIMENT FROM A COLLEAGUE

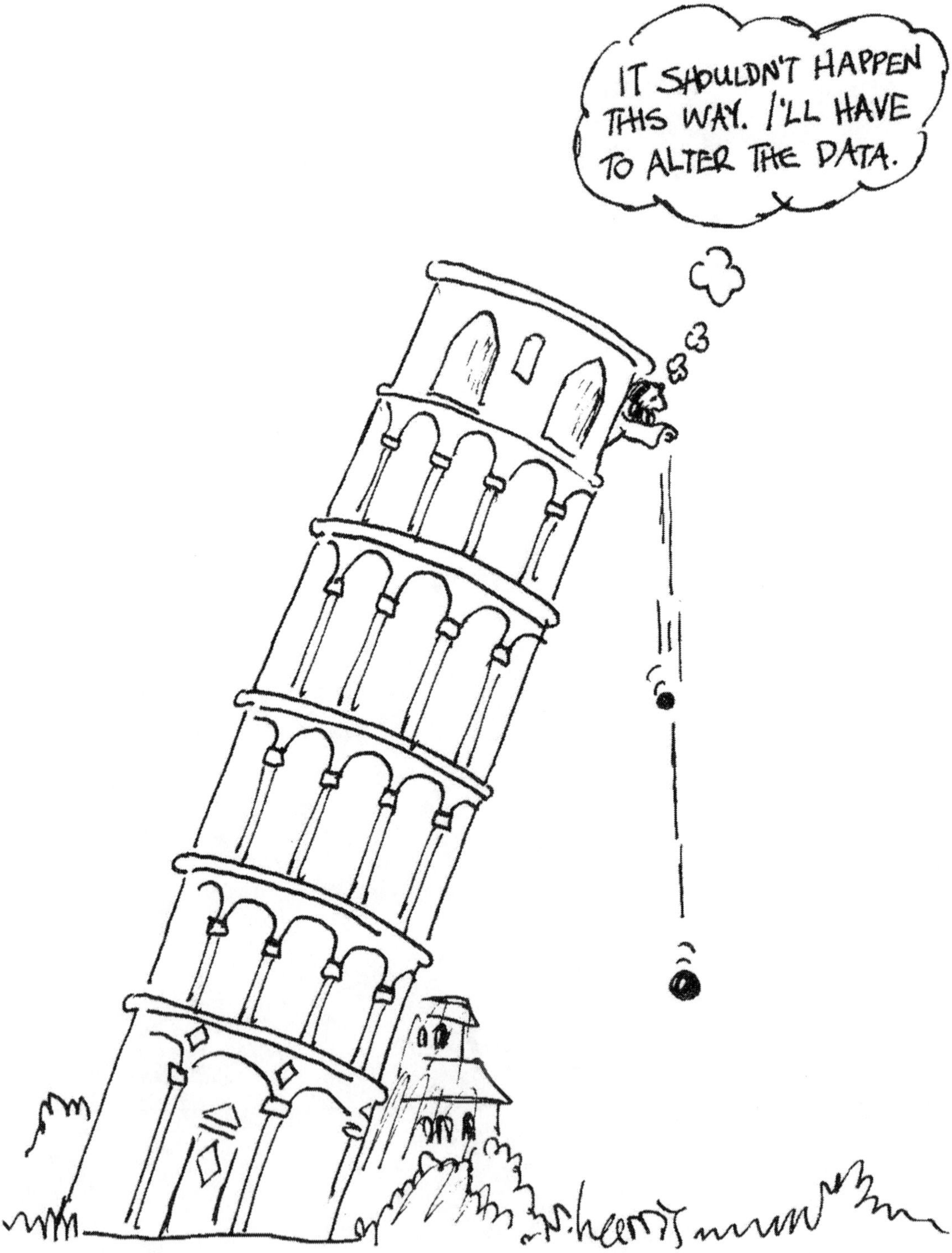
IT SHOULDN'T HAPPEN THIS WAY. I'LL HAVE TO ALTER THE DATA.

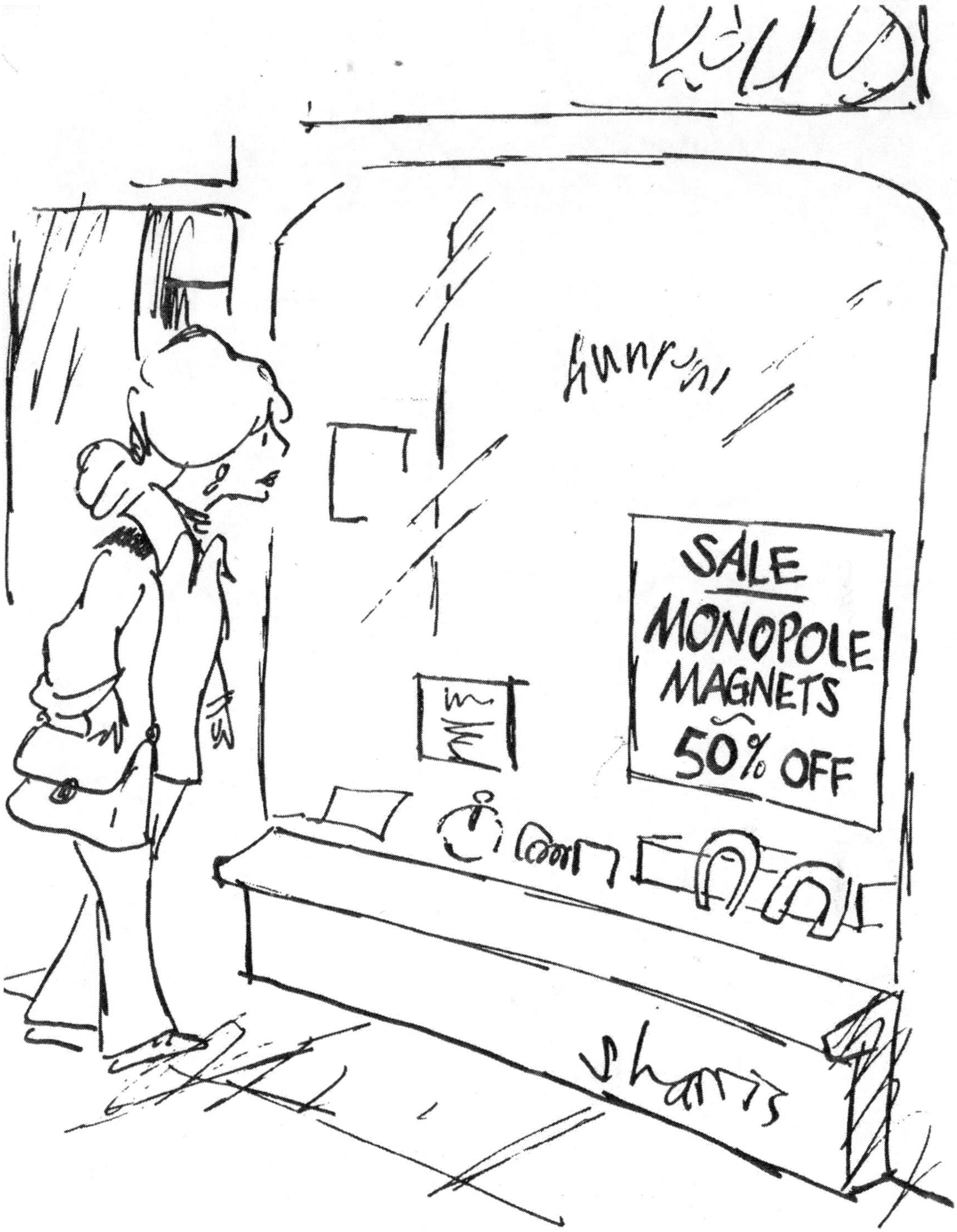
SALE
MONOPOLE
MAGNETS
50% OFF

"THERE GOES ANOTHER ONE. GALILEO, DON'T YOU HAVE _ANY_ IDEA WHAT MAKES THEM DO THAT?"

THE VENN-DIAGRAM BUILDINGS

BIOLOGY
+ CHEMISTRY & PHYSICS
& A LITTLE PHYSICAL CHEMISTRY

CHEMISTRY
+ BIOLOGY & PHYSICS
& A LITTLE BIOPHYSICS

PHYSICS
+ BIOLOGY & CHEMISTRY
& A LITTLE BIOCHEMISTRY

S.Harris

AS SMART AS HE WAS, ALBERT EINSTEIN COULD NOT FIGURE OUT HOW TO HANDLE THOSE TRICKY BOUNCES AT THIRD BASE.

"IF I'VE LEARNED ONE THING IN MY LONG REIGN, IT'S THAT HEAT RISES."

"I THINK YOU SHOULD BE MORE EXPLICIT HERE IN STEP TWO."

"IT'S DOWN. THE PARTICLES ARE SUPPOSED TO COLLIDE WITH OTHER PARTICLES, NOT WITH THE PHYSICISTS."

"THERE GOES ARCHIMEDES WITH HIS CONFOUNDED LEVER AGAIN!"

"SOMETIMES SHE'S LIKE A WAVE, SOMETIMES SHE'S LIKE A PARTICLE."

"IF THIS IS CORRECT, THEN EVERYTHING WE THOUGHT WAS 'A WAVE' IS REALLY A PARTICLE, AND EVERYTHING WE THOUGHT WAS 'A PARTICLE' IS REALLY A WAVE."

"WHAT IF WE SPEND ALL THESE BILLIONS, AND THERE JUST _AREN'T_ ANY MORE PARTICLES TO FIND?"

"COBALT? TITANIUM? WHAT I'M LOOKING FOR IS LITHIUM, COLUMBIAN AND YTTRIUM."

"I'VE DISCOVERED WHAT I BELIEVE IS THE ELEMENTARY
PARTICLE: A SMALL STONE."

"WHAT AN EGO! ONE DAY IT'S NEWTON'S LAWS OF DYNAMICS, THEN IT'S NEWTON'S THEORY OF GRAVITATION, AND NEWTON'S LAW OF HYDRONAMIC RESISTANCE, AND NEWTON'S THIS AND NEWTON'S THAT."

원자화된 아인슈타인

쿼크박사 겉으로 보면 아인슈타인이 보이지만 자세히 들여다보면 점들만 보인다. 일상의 물질들은 분자들로 이루어졌고 분자는 원자들이 만들며 원자는 원자핵과 전자로 만들어졌으며 양성자와 중성자는 쿼크들로 만들어졌다. 따라서 우주의 가장 작은 알갱이들은 전자, 쿼크들이다. 이런 분야의 연구는 고에너지 물리학 또는 입자물리학이라 불리는 분야에서 이루어진다.

EINSTEIN ATOMIZED

"AFTER YEARS IN THEORETICAL PHYSICS, I CAN'T GET ENOUGH EQUIPMENT TO PUT MY HANDS ON."

TO AN OBSERVER APPROACHING
THE SPEED OF LIGHT, EINSTEIN AND
HIS SURROUNDINGS APPEAR TO BE
TALL AND THIN

"MY GOODNESS, IT'S 12:15:0936420175! TIME FOR LUNCH."

"WE SEEM TO BE AT THAT POINT WHERE PARTICLE PHYSICS LEAVES OFF AND THEOLOGY BEGINS."

1.

2.

3.

4.

"IF TACHYONS DO EXIST, AND IF THEY DO GO FASTER THAN THE SPEED OF LIGHT, THEN I'M DETERMINED TO FIND SOMETHING THAT GOES FASTER THAN TACHYONS."

"THERE MUST BE SOME WAY WE CAN CAPITALIZE ON THAT DAMN BOSON."

"ITS TOP SPEED IS 186 M.P.H. THAT'S $\frac{1}{3,600,000}$ THE SPEED OF LIGHT."

They gave me the Gauss theorem, the Gauss method for quadratics, Gaussian distribution, the Gauss eyepiece...
And I've got the Newton laws of dynamics, Newton's formula for interpolation, Newton's laws of cooling, Newton's rings...
J. Harris

CONDENSED-MATTER BLUES

THINGS ARE SPREADIN' OUT,
GETTIN' TOO DIVERSE TO SEE.
YES, IT'S ALL SPREADIN' OUT,
TOO DIVERSE FOR YOU 'N' ME.

'CAUSE I'M WORKIN' WITH
CONDENSED MATTER,
   DON'T WANT IT TO SPREAD
      OR SPLATTER,

   AN' I GOT THOSE CONDENSED-
   MATTER BLUES...

"PARTICLES, PARTICLES, PARTICLES."

"ENTROPY ALREADY?
YOU JUST MADE IT."

"I'M SO RELIEVED NEW IDEAS HAVE COME UP. I COULDN'T MAKE HEAD OR TAIL OF STRING THEORY."

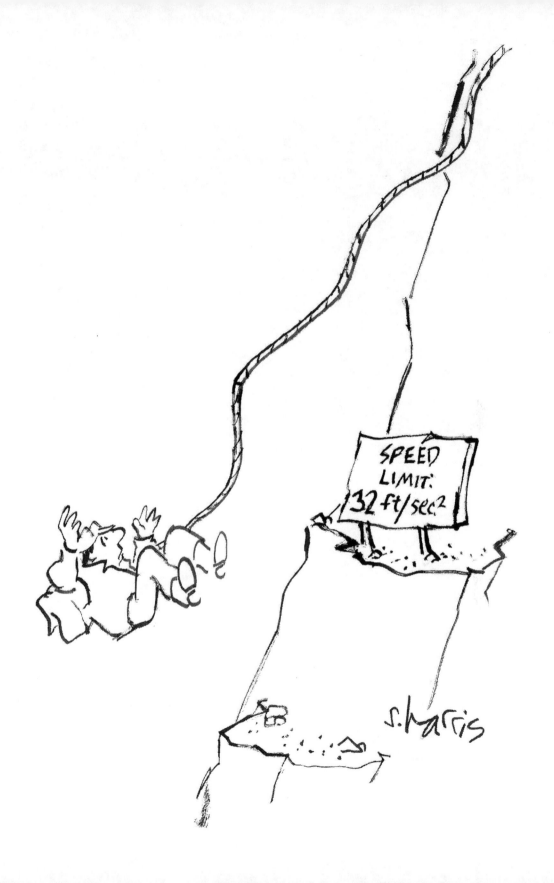

"WHAT'S COME OVER HEISENBERG? HE SEEMS TO BE CERTAIN ABOUT EVERYTHING THESE DAYS."

"IS THAT IT? IS THAT
THE BIG BANG?"

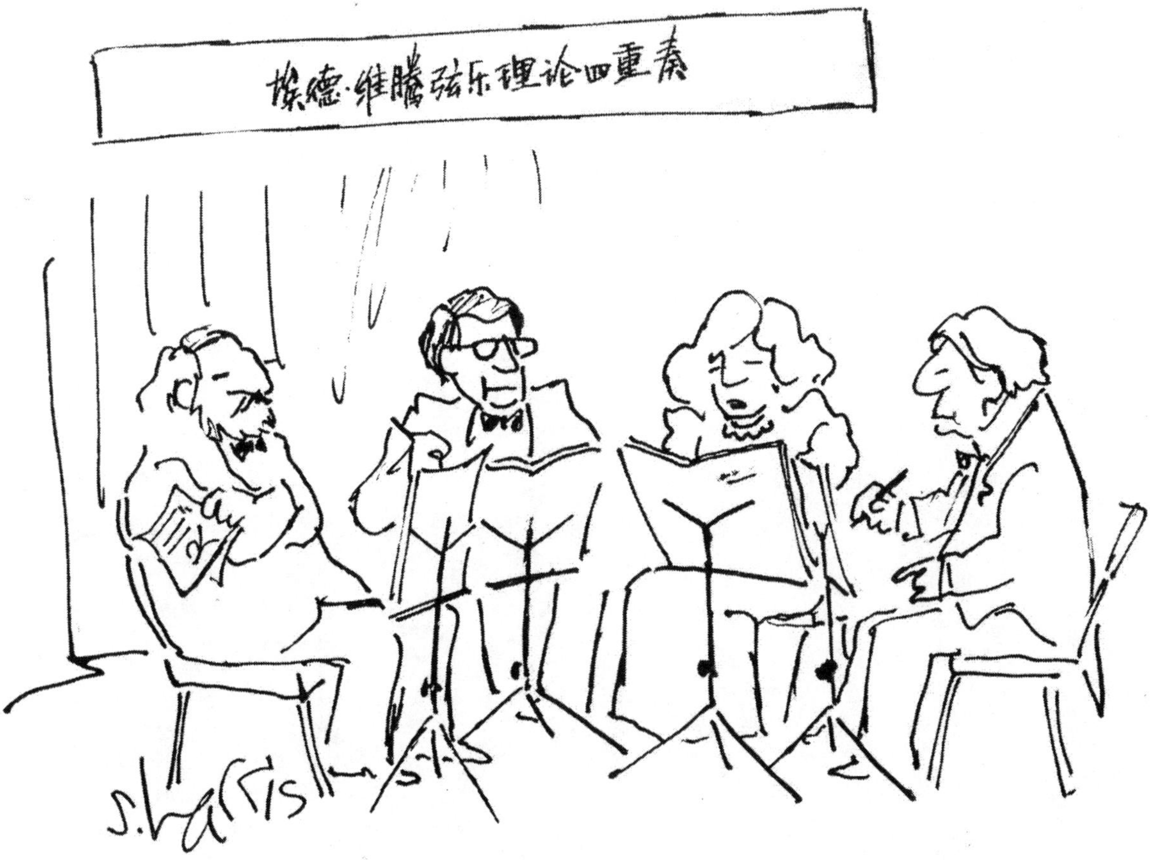
埃德·维滕弦乐理论四重奏
S. Harris

"THE SMALLER WE MAKE 'EM, THE BIGGER WE GET."

"BUT IF THE PARTICLE IS TOO SMALL AND TOO SHORT-LIVED TO DETECT, WE CAN'T JUST TAKE IT ON FAITH THAT YOU'VE DISCOVERED IT."

"DON'T MIND HIM. AS WE TAKE OUT THE COAL, HE FILLS THE SPACES WITH NUCLEAR WASTE."

"IT'S UNIFIED AND IT'S A THEORY, BUT IT'S NOT THE UNIFIED THEORY WE'VE ALL BEEN LOOKING FOR."

"BUT YOU CAN'T GO THROUGH LIFE APPLYING HEISENBERG'S UNCERTAINTY PRINCIPLE TO EVERYTHING."

PHYSICISTS AT THE FOURTH GROTSCHLOV CONFERENCE ASSEMBLED TO DETERMINE ONCE AND FOR ALL IF LIGHT IS A WAVE OR A PARTICLE

"I LOVE HEARING THAT LONESOME WAIL OF THE TRAIN WHISTLE AS THE MAGNITUDE OF THE FREQUENCY OF THE WAVE CHANGES DUE TO THE DOPPLER EFFECT."

"OF COURSE YOU CAN'T PUT YOUR FINGER ON IT. IT'S A HYPOTHETICAL PARTICLE."

"ONE HUNDRED MILLION NEUTRINOS ARE PASSING THROUGH OUR BODIES EVERY SECOND, AND WE'RE WORRIED ABOUT THE PRICE OF COFFEE."

"...BUT HERE IN THE OUTSIDE WORLD, I NEED MORE TIME AND I NEED MORE SPACE."

"AH—PHYSICS, COMING UP THE RIVER FROM NEW ORLEANS."

COUNTING GEIGERS...

1. ELIZABETH 2. HANS 3. MARGOT 4. FRIEDRICH
5. SOPHIA 6. WILHELM 7. CORETTA 8. PIETER 9. SIGMUND

"GOOD NIGHT, AL. WE STILL HAVEN'T FOUND IT."

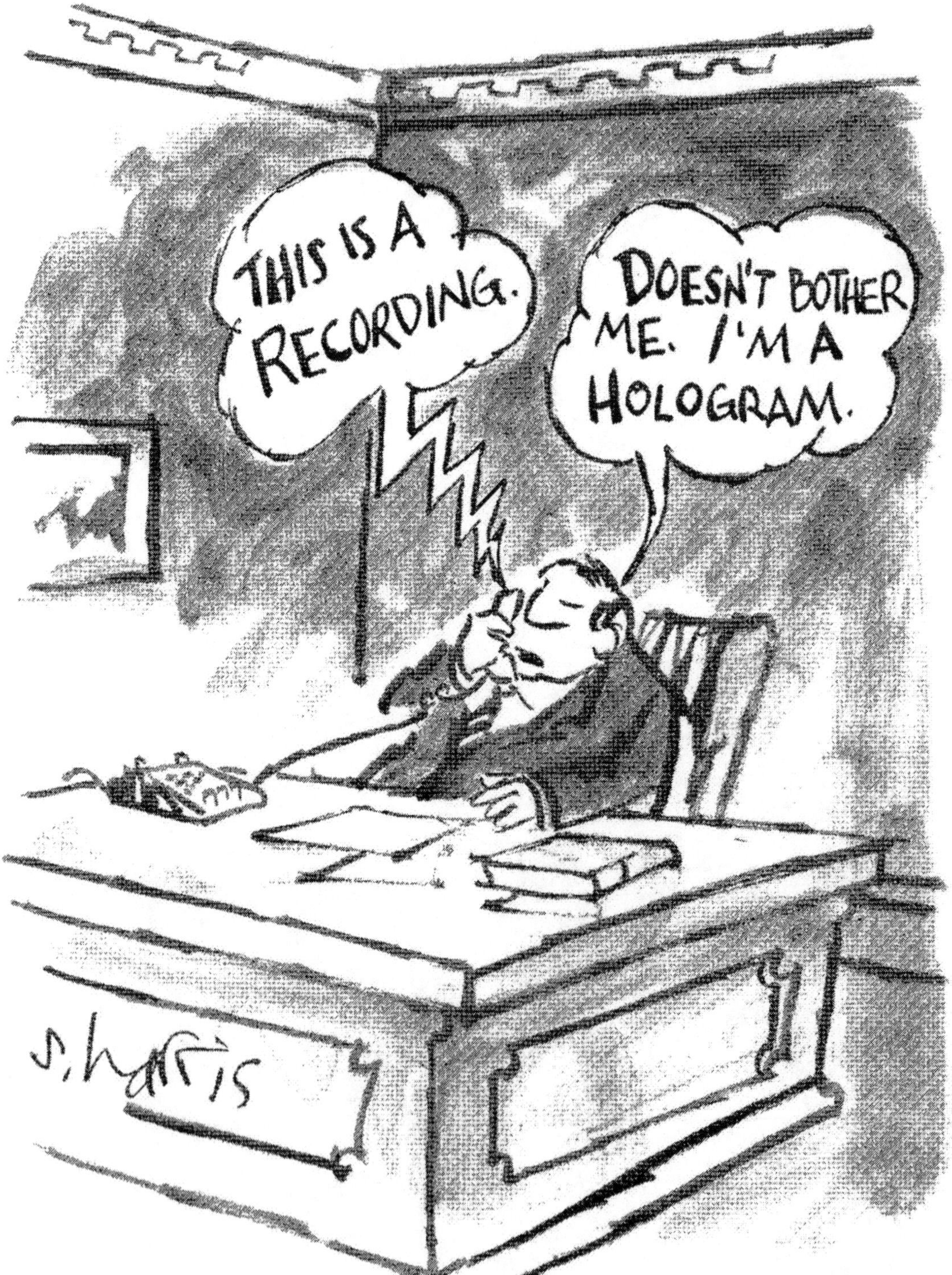
THIS IS A RECORDING.
DOESN'T BOTHER ME. I'M A HOLOGRAM.
S. HARRIS

"A Ph.D. IN PARTICLE PHYSICS, EXPERIENCE IN AEROSPACE AND ROCKETRY... OF COURSE I CAN JUGGLE."

ARCHIMEDES
GALILEO
COPERNICUS
Καταρανένο μήλο!*
DANNATA MELA!*
Cholernie* jabłko!
J. Harris
* DAMN APPLE

Made in United States
Troutdale, OR
08/27/2023

12395378R00086